P·I·C·T·U·R·E·P·E·D·I·A

NOTE TO PARENTS

This book is part of PICTUREPEDIA, a completely
new kind of information series for children.
Its unique combination of pictures and words
encourages children to use their eyes to discover and
explore the world, while introducing them to a wealth
of basic knowledge. Clear, straightforward text
explains each picture thoroughly and provides
additional information about the topic.

'Looking it up' becomes an easy task with
PICTUREPEDIA, an ideal first reference for all types of
schoolwork. Because PICTUREPEDIA is also entertaining,
children will enjoy reading its words and looking
at its pictures over and over again. You can encourage
and stimulate further inquiry by helping your child
pose simple questions for the whole family to
'look up' and answer together.

MAMMALS

DORLING KINDERSLEY

LONDON, NEW YORK, AUCKLAND
DELHI, MUNICH, SYDNEY

DK www.dk.com

First published in Great Britain in 1993
by Dorling Kindersley Limited, London

This updated edition published in 2000 by:

Dorling Kindersley Limited
9 Henrietta Street, London WC2E 8PS, Great Britain

Dorling Kindersley Publishing Pty Limited
(A.C.N. 078 414 445)
118-120 Pacific Highway, St Leonards NSW 2065, Australia

Dorling Kindersley (India) Pvt. Ltd.
102/3 Kaushalya Park, Hauz Khas, New Delhi 110016, India

A CIP catalogue record for this
book is available from the British Library.

ISBN 0 7513 6910 1

Reproduction by Colourscan, Singapore
Printed and bound by L. Rex Printing Company Limited, China

MAMMALS

A DORLING KINDERSLEY BOOK

CONTENTS

WHAT IS A MAMMAL?

Mammals are amazing animals. Some climb through trees, others race across the ground, burrow, swim or even fly. They come in many shapes and sizes too. A giraffe is tall, a mouse is small and a platypus looks like an otter with a duck's beak. So what makes them all mammals? They are hairy, and feed their babies milk. You are a mammal too.

Small Start
A few furry animals scurried between the feet of dinosaurs 195 million years ago.

Warm Inside
This arctic hare is sitting on icy snow and the mongoose on hot desert sand. But the temperature inside their bodies is almost the same. This is because mammals are warm-blooded. They can warm up and cool down their bodies.

Giraffe
(female)

Asian elephant
(male)

Tiger

Giant
otter

Gelada
(male)

Koala

Royal
antelope

Desert
hedgehog

Odd-Mammals-Out

Most mammals are born, not hatched from eggs. Only two types of mammal lay eggs: platypuses and hedgehog-like echidnas. Platypus eggs hatch after just ten days. The furry, beaked babies are then fed on milk for four months.

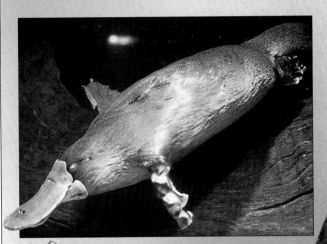

This is an ocelot. It is one of more than 4,000 species of mammal.

Mammals are the only animals to have ear flaps.

All mammals breathe air into lungs.

The lower jaw is made up of one piece of bone.

Platypus eggs are soft and the size of marbles.

All mammals have hair – even whales have hairy lips!

Ocelots are in danger of disappearing for ever because they are killed for their fur coats. More than 550 mammal species are threatened with extinction.

Female ocelots give birth, after a 70-day pregnancy, to up to four kittens.

Giant panda

Little brown bat

Inside the body there is a frame of bones, called a skeleton.

Fat keeps in the body's heat.

Black-tailed jack rabbit

Baby Food

Female mammals make milk in their mammary glands. With nothing to do but sleep and suck milk, or suckle, their babies grow quickly.

Golden mice

MARSUPIALS

A marsupial is an animal that has a pocket, called a pouch, on its tummy for carrying its babies. Inside this nursery the baby is safe and has milk to drink. Today, most mammals with pouches live in Australia, but 50 million years ago they lived in the Americas and in Antarctica. Most marsupials died out when the more modern mammals, such as horses, cats and rats developed. Marsupials survived in Australia because the 'new' mammals could not reach the island. Kangaroos are the most famous marsupials but there are also marsupial 'mice' and 'dogs'.

Australia

Female red kangaroos are one metre tall. Males are twice as big. They live in Australia.

Kangaroos can't walk backwards!

Like deer, kangaroos have long faces to make room for their big, flat, grass-grinding teeth.

Female red kangaroos are called blue fliers because they have blue-grey fur and bounce faster than the red males.

The tail helps it to balance as it bounds along.

Kangaroos lick a bald spot on their arms to cool down! As the saliva dries it takes heat away.

Only females have pouches – males don't need them because they don't have babies!

The baby, or joey, hops into the pouch if it sees an eagle or dingo.

Jump to It!
A kangaroo's back legs are so big that it would fall over if it ran. But they are good for jumping - a red kangaroo can bounce along at 65 kilometres per hour.

Huge leg muscles

8

Missing Marsupial

The last Tasmanian wolf is thought to have died in a zoo in 1936. It was striped like a tiger and had a thick tail like a kangaroo. Farmers shot them all because they ate sheep.

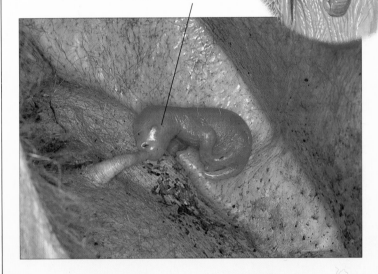

Birth

Jumping Bean?

A newborn wallaby looks like a red bean! It is less than two centimetres long and has no legs, hair or eyes. Like all marsupials, it carries on growing in a pouch, not inside the mother's body.

The 'bean' squirms through the forest of hair by waving its stumpy arms.

Three minutes later it reaches the pouch.

It hooks onto a nipple and starts to suck milk.

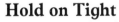

Acting Star

Opossums are the only living American marsupials. When a Virginia opossum is attacked it sticks out its tongue, lies very still and pretends to be dead – it plays possum!

Hold on Tight

This baby koala has grown too big to fit inside its mother's pouch, so it clings to her fur as she clambers through the eucalyptus leaves!

Mammals with Pouches

Tasmanian devil

Honey possum

Numbat

Ring-tailed rock wallaby

TOOTHLESS MAMMALS

Anteaters and pangolins do not have teeth at all. Armadillos and aardvarks have small 'pegs' at the back of their mouth, but no big biting teeth – unlike you, they could not bite into a crunchy apple. Tough teeth are just not needed for licking up ants and termites. Sloths slowly chew leaves and don't have any proper teeth either. Thirty-seven species of these wonderful but weird mammals still survive, mainly in South America.

Ant

Ant Scratchings
A pangolin can't get its claws between its scales to scratch. So this 'walking pine cone' lets ants climb over its body and eat the insects that cause the itch. When it wants to wash off the ants, it wades into a river. Then it eats the drowned ants!

Tough plates of horn protect the soft body.

Naked-tailed armadillos are 15 centimetres tall. They live in Guatemala, Belize and Venezuela.

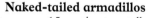

Armadillos sniff out ants and termites.

This armadillo escapes from danger by quickly digging a tunnel with its axe-like claws. Some armadillos roll into a ball instead!

Sticky Lick
An anteater's tongue is as long as your arm and covered in sticky saliva.

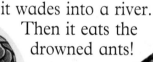

Anteaters eat more than 30,000 ants a day!

Creepy-Crawly Coat
A sloth is slow but its coat is full of life! Moths live on its hairy coat. Caterpillars that hatch from the moths' eggs eat tiny plants that grow in grooves in the sloth's hairs.

Animal Jigsaw
An aardvark has a body like a pig, a tail like a kangaroo, ears like a rabbit and it licks up ants and termites like an anteater. But it is not related to any of these other mammals.

Go Slow Sloth
Three-toed sloths are the slowest mammals in the world. When they drag themselves along the ground they can only travel two metres a minute, half the speed of a tortoise!

Mammals that Can't Bite

Nine-banded armadillo

Giant anteater

Giant pangolin

Two-toed sloth

It gulps air into its stomach to help it float across a river!

The plates are linked by leathery skin. The armadillo could not bend its body without these joints in its suit of armour.

This is the only species of armadillo not to have an armoured tail.

This part is not armour-plated. It is covered in soft hairs.

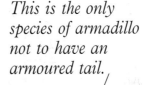

When this naked-tailed armadillo is older it will lose its pink colour and turn almost black.

INSECTIVORES

Insectivores are sharp-toothed, long-nosed animals that munch insects and juicy worms, slugs and snails! Their busy little bodies lose heat easily, so they need to eat frequently. The food they eat produces the energy needed to keep them warm. But how do insectivores survive winters, when there are fewer insects to eat? Shrews search through rotting leaves and most manage to find enough food. Moles stay underground and hedgehogs spend cold winters in a deep sleep, called hibernation.

How Hungry?
Imagine having to eat a pile of food that weighs twice as much you – shrews have to do this every day!

A Bite for Lunch
The water shrew is one of the few poisonous mammals. Its saliva can kill frogs, but not people!

The tiny eyes are covered by fur. A mole can only just tell the difference between light and dark.

Little bumps on its tail and its nose help this European mole sense where it is going.

A mole's wide front feet are shaped like spades – ideal for digging.

Molehill

The grass nest is the size of a football.

Worms burrow into the tunnel and are caught by the mole.

Moles turn round by doing forward rolls. If the tunnel is too narrow, they run backwards.

Moles live alone. This worm-thief will soon be chased away.

Moles eat more than 50 worms a day! Live ones are stored in a larder.

Greedy Guts

Shrews often eat animals that are bigger than themselves. This long, juicy worm will fill its tummy for two or three hours!

Worms in a Week

Streaked tenrecs grow up faster than any other mammals. They stop drinking milk and start to eat worms when they are only six days old.

Insect-Eaters

Golden mole

A hedgehog can stay rolled up for hours.

Head

Roll Up, Roll Up!

Fearless hedgehogs don't run away from danger – they stick out their spines and roll into a ball. No one wants to eat a mouthful of needles!

Some foxes and badgers have learned to push hedgehogs into puddles to make them unroll!

Desman

Star-nosed mole

The hedgehog's skin is larger than its body. When it curls up, it can pull its prickly skin over its head!

Spines are just stiff, hollow hairs.

One-week-old baby shrews hold on to one another so that they don't get lost.

Adult European hedgehogs have more than 5,000 needle-sharp spines.

Solenodon

White-tailed shrews

WEASELS AND MONGOOSES

Many mammals eat meat. Those that have special teeth for killing and eating other animals are known as carnivores. Long front teeth, called canines, are used like daggers to stab, and scissor-like back teeth slice meat off bones. Although tigers and wolves are huge, most carnivores are in fact small. Weasels, skunks, and their relatives, the mongooses, are fierce – they rip their prey apart in a flurry of fur and teeth.

Badger's canine

Tiny but Tough
The American least weasel is the smallest carnivore in the world. It weighs about as much as ten sugar lumps.

A badger's sense of smell is more than 700 times better than yours!

Black and white stripes look like shadows in the dark and make badgers hard to spot.

Like all members of the weasel family, badgers have long, tube-shaped bodies – ideal for moving through tunnels.

Ear

Badgers' wiry hair used to be made into bristles for brushes.

Slugs and rats will do, but worms are a badger's favourite food – they eat hundreds each night!

Fur Coats for Mink
Many mammals are killed for their furry coats. Today there are not enough mink left in the wild to trap, so they are bred in fur farms.

The powerful jaws can crunch big bones.

Short legs

Eurasian badgers are 30 centimetres tall. They live in Europe and Asia.

Badgers sharpen their big, digging claws on trees.

One bite from a cobra's fangs can kill a mongoose.

The mongoose bites the cobra's head to kill it.

Even when it is knocked off its feet, the mongoose does not let go.

Teeth Beat Fangs
A brave, swift mongoose can defeat a cobra! While the snake lifts up its head to spit its poison, the mongoose darts in and clamps its canine teeth around the snake's neck.

Big Squirt
All small carnivores smell, but a stripy skunk stinks! When it is threatened it lifts up its tail and squirts out two jets of horribly smelly liquid.

Meerkats on the Menu!
Many small carnivores are eaten by bigger meat-eaters. Meerkats, a type of mongoose, eat scorpions and snakes but they have to watch out for foxes or they themselves will be eaten.

Better Safe than Sorry
Giant otters in the muddy Amazon river eat most fish they find head first, but they leave the biting jaws of the piranha till last!

This tiny mouse-killer can squeeze through a ring!

Small Carnivores

Zorilla

American least weasel

Banded palm civet

African linsang

CATS

Cats are carnivores. Most creep up on their prey by sneaking slowly and silently through the undergrowth. Then, suddenly, they will hurl themselves onto their surprised victim. The sharp canine teeth quickly deal the deadly blow. The biggest cat of all, the stripy tiger, can eat 25 kilograms of meat in a meal! But this terrifying animal never meets the lion, the king of the cats, because lions live in Africa and tigers in Asia.

A cheetah can accelerate as quickly as a Ferrari car.

Tigers

Bright Eyes
When light shines on a cat's eyes, they glow like the reflectors on the back of a bike. This happens because the light bounces back off a special layer in the cat's eyes. This layer collects light. It helps cats see six times better than you in dim light.

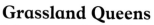

Aerial Ambush
All cats climb trees. This spotty jaguar is waiting to drop down on a passing peccary or tapir. It will even tackle giant alligators.

Grassland Queens
Lions are the odd-cats-out because they live in groups, called prides. The male lions are often called the 'Kings of the Jungle', but in fact they do not live in jungles and the females, or lionesses, are in charge.

Bendy backbone

A cheetah can only sprint for 20 seconds because it soon tires.

Speedy Cat

Cheetahs are the only cats that run down their meals. They can sprint at 100 kilometres per hour and are the fastest mammals.

Play Fight

Baby cats, or kittens, are playful. They chase one another, leap into the air and chew twitching tails! This is the way they learn to hunt.

Sensitive whiskers

Leopards are 60 centimetres tall. They live in Africa and southern Asia.

Leopards sleep 16 hours a day, usually in short 'cat naps'.

The rough tongue can rub meat off bones!

Big cats roar; they can't purr.

Really Wild Cat

There are more than 300 million pet cats in the world. They are believed to have been bred from wildcats over 3,000 years ago. This Scottish wildcat is much fiercer than a tame tabby!

Black leopards, or panthers, can be born to 'yellow' parents.

Panthers are spotty, but their spots are hard to see!

Leopards live and hunt on their own. Groups of lions often gang up on them and steal their food.

Cats comb and clean their fur coat with their tongue.

Soft pads let cats creep quietly.

Put your Claws In!

Little muscles pull most cats' dagger-like claws into special pockets, to stop them from becoming blunt.

DOGS

Like all carnivores, dogs eat meat. On its own a dog can only trap animals that are smaller than itself, but 20 African hunting dogs, working together, can easily catch and kill a zebra. Most dogs have learned this lesson and prefer to hunt in family groups, called packs. The 35 types of wild dogs have often been treated as enemies, not loved like pet dogs. Wolves have been wiped out in many places. Foxes only survive because they are smaller and more cunning.

All pet dogs have been bred from grey, or timber, wolves. The first dogs were tamed more than 12,000 years ago!

Hooooowl!
In the dead of night, the wolves in a pack get together, throw back their heads and howl. This warns other wolves to keep out of their territory.

A wolf can hear a watch ticking over ten metres away.

Grey wolves are about one metre tall. They live in Canada, the USA, Europe and northern Asia.

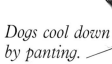

Dogs cool down by panting.

Pointed canine teeth stab the prey. Cheek teeth slice the meat into pieces that are small enough to swallow.

Grey wolves trot more than 60 kilometres a day when they are hunting a moose or an ox.

Pack Property
Female African hunting dogs can have as many as 16 babies. These puppies belong to the whole pack, not just the mother. They are even suckled by other females.

This pup is eight weeks old. It already eats meat but will not go hunting with the pack until it is six months old.

New Neighbour

Red foxes used to live in woods but many have moved into cities. They scamper through the streets at night searching for fruit and mice, or bins to raid!

Cleaning Up

Big African dogs, called jackals, love leftover lion-food – lions hardly ever finish their dinners! Carnivores that do not kill their own food are called scavengers.

A fox's tail is called a brush.

A grey wolf's thick coat can be any colour from white to black!

Bushy tail

Dog 'Talk'

Every dog has to know its place in the pack – they can't all be the leader! Dogs can't talk, so they use body language instead to let one another know whether they give or take orders.

Win by a Nose

When you smell a flower, you can often tell what sort of flower it is, without opening your eyes. Dogs can do much better than this – they can smell who touched the flower the day before!

Dogs wag their tails when they are happy.

Dogs are marathon runners, not sprinters. A wolf can only run at 45 kilometres per hour – much slower than a lion.

The pack leader holds its tail upright and snarls.

Ankle

This dhole does not want to argue, so it rolls on its back.

Dogs walk on their toes.

The claws stay out all the time.

BEARS

Bears are big and usually have thick, shaggy coats. Brown bears are the most common but giant pandas are more famous. People have argued for years about whether giant pandas are bears or not. Scientists now think, as any child can see, that they are black-and-white bears! Bears look cuddly, but they are fierce. People have shot so many of these big beasts that today bears only survive in hilly hideaways.

The Big Sleep
Bears that live in cold places spend the winter inside warm caves. The females give birth to their tiny cubs, usually twins, while they are asleep.

Polar bear pawprint

Honey and Grass for Tea?
Most bears eat all sorts of things – they are omnivores. These are a few of their favourite foods.

Honey

Grass

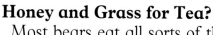

Berries

Masters of Disguise
Polar bears live in the icy Arctic and are the only completely carnivorous bears. Seal blubber is their favourite food.

Open Wide!
A grizzly bear has a simple way of fishing. It stands in a river and waits for fish to leap right into its mouth.

The polar bear hides its black nose with its white paw.

It sneaks up on the seal by pretending to be an iceberg!

The cunning bear springs out of the icy water to kill the surprised seal with one swipe of its huge paw.

Half-webbed toes

Ringed seal pup

Pawprints

Unlike cats and dogs, bears have flat feet. Their heels touch the ground when they walk.

A special pad on a panda's paw is used as a sort of thumb – useful for grasping bamboo shoots.

All bears have small, round ears.

This big brown bear is over twice the size of a tiger!

A bear's face always looks the same - you can't tell whether it is angry or happy!

When bears get an itchy back they rub it against a tree!

Save the Giant Panda!

There are less than 1,000 giant pandas left. It is not going to be easy to save them – females are only fertile for three days a year. They also need to eat 20 kilograms of one special type of bamboo a day.

Many bears have rotten teeth. This is because they love sweet foods – especially honey!

With their big, strong arms, bears can hug a person to death!

Grizzly bears are a type of brown bear. They are called grizzly bears because the tips of their brown hairs are grey, or grizzled.

Grizzly bears stand up to 3 metres tall. They live in Canada and the USA.

The front paws can be used as clubs to hit other large animals.

21

APES

Playful   Begging for food

There are four sorts of ape: chimpanzees, orangutans, gibbons and gorillas. They all live for many years, have big brains, no tails and can walk upright. Apes are the closest relatives of people.

Orangutan (male)

Gorillas are the biggest and strongest apes but they are gentle giants. Chattering chimps are cheeky and cute but much more dangerous – they even kill deer and monkeys to eat! Family life is important to all these intelligent animals: chimps cuddle and shake hands when they meet.

Playtime
Baby chimps take a long time to grow up. Their mothers feed them milk for five or six years – so they have plenty of time to play.

Brainy Beast
Chimps are one of the few animals to use tools. They use leaves as sponges! First they soften handfuls of leaves by chewing them and then they use them to soak up water.

Go Bananas
Gorillas really do eat bananas. They also like nettles, giant celery and banana leaves.

Walk Like an Ape
All apes can stand on just two feet, but they usually walk on all fours like this.

Apes walk on their knuckles.

Not a Word
The chimp is one of the few mammals that can pull faces to show its feelings.

Frightened   Angry

Holding Hands
You can pick things up because you are able to fold your thumb across your hand. Apes and monkeys have these useful, 'opposable' thumbs too.

Like you, apes have sensitive hands.

Male gorillas stand 1.7 metres tall. They live in Africa.

An orangutan's big toes can grip things too!

Gorillas have about the same number of hairs as you. They look more hairy because it is long and grows all over their body.

Big brain

Gorillas can climb trees but they spend most of the day lazing on the ground.

Is it a Bird?
Every night, orangs build a cosy nest to sleep in. It takes just five minutes to build a mattress of branches and a blanket of leaves.

Apes see things in colour – just like you.

Baby gorillas learn to crawl at ten weeks, climb at five months and walk at eight months. They may live to be forty.

All apes can sit and stand up straight.

MONKEYS

Monkeys are primates. It is easy to tell them apart from the cleverer primates, people and apes, because they have tails. Some, such as mandrills, live on the ground but most monkeys are light enough to jump or swing through the trees. They always look before they leap though, because there is danger all around. Large eagles may swoop down from above and leopards lurk below. If they miss their footing, monkeys may plunge up to 60 metres to the ground – like falling from the 13th floor of a building!

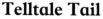
Squirrel monkey

Built to Balance
If you start to lose your balance, you can use your arms to steady yourself. Monkeys use their tail instead – leaving their arms free for climbing

Keep it Clean!
These rhesus monkeys are queuing up to have insects and dirt picked out of their fur. They even pick one another's teeth clean! This grooming helps to keep them tidy and also good friends.

Groups of monkeys are called troops.

Face to Face

Telltale Tail
Ring-tailed lemurs are primitive primates and live in troops like monkeys. They keep together in tall grass by pointing their tails upwards.

This lemur is looking for a tail to follow!

Mandrill (male)

Proboscis monkey (male)

Bald uakari

Cotton-top tamarin

A Gripping Tail

Many South American monkeys have three 'arms' – their tails are prehensile, so they can hold on to things. A spider monkey can hang by its strong tail, leaving both hands free for feeding.

Both eyes face forwards to spot safe landing places!

All primates, including you, see in colour.

Narrow chest

Mane of soft, silky hair.

Monkeys are not fussy eaters. They eat fruit, flowers, lizards, butterflies and even frogs' legs!

Monkeys' legs are shorter than their arms.

This tamarin weighs only 600 grams, so it is light enough to scamper across small branches without breaking them.

People have chopped down so many of the trees that golden lion tamarins live in that there are fewer than 200 of these beautiful monkeys left.

Hairy tail

Golden lion tamarins are 22 centimetres tall. They live in Brazil.

Bathtime

Japanese snow monkeys never get cold feet. They spend much of the winter sitting in pools of hot water that bubble up from beneath the icy ground.

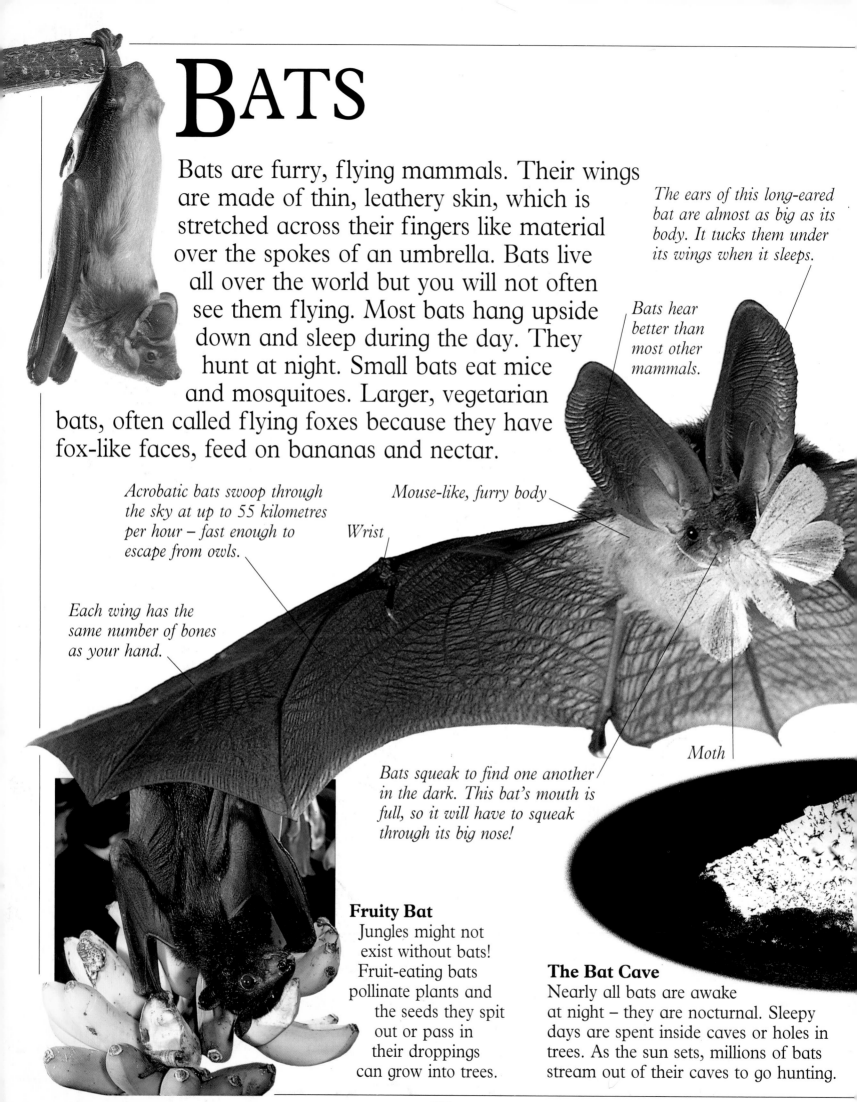

BATS

Bats are furry, flying mammals. Their wings are made of thin, leathery skin, which is stretched across their fingers like material over the spokes of an umbrella. Bats live all over the world but you will not often see them flying. Most bats hang upside down and sleep during the day. They hunt at night. Small bats eat mice and mosquitoes. Larger, vegetarian bats, often called flying foxes because they have fox-like faces, feed on bananas and nectar.

The ears of this long-eared bat are almost as big as its body. It tucks them under its wings when it sleeps.

Bats hear better than most other mammals.

Acrobatic bats swoop through the sky at up to 55 kilometres per hour – fast enough to escape from owls.

Mouse-like, furry body

Wrist

Each wing has the same number of bones as your hand.

Moth

Bats squeak to find one another in the dark. This bat's mouth is full, so it will have to squeak through its big nose!

Fruity Bat
Jungles might not exist without bats! Fruit-eating bats pollinate plants and the seeds they spit out or pass in their droppings can grow into trees.

The Bat Cave
Nearly all bats are awake at night – they are nocturnal. Sleepy days are spent inside caves or holes in trees. As the sun sets, millions of bats stream out of their caves to go hunting.

Camping Bats

Imagine having to build a new house every night – tent bats do! These small, white bats nibble through the middle rib of a palm leaf until it droops down to form a tiny tent. The bats hang underneath, out of the wind and rain.

The Smallest Mammal

The hog-nosed bat can fit in the palm of your hand. Its body is three centimetres long and it has a wingspan of just 15 centimetres. It weighs less than one grape.

This is a thumb. Bats use their thumbs as combs to groom their fur and as hooks to hold on to things.

The wing can be used as a scoop to catch flying insects.

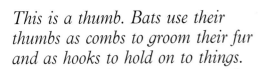

Long finger

Long-eared bats have a wingspan of 28 centimetres. They live in northern Europe.

Fangs for Dinner!

Vampire bats love the taste of blood. This one has sliced open the foot of a sleeping chicken with its sharp teeth. It will lap up about one tablespoonful of blood.

Sounds Tasty

Many bats don't use their eyes to see, they use their voice and ears instead! American fishing bats make clicking noises as they fly over ponds. When these sounds bounce back off ripples, they know that a fish is near to the surface.

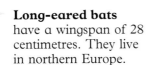

The bat hears the tiny echo and swiftly swoops down to grab the fish.

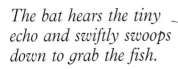

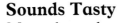

SMALL RODENTS

These newborn, rubbery, wriggling mice can only squeak, sleep and suckle.

A rodent is an animal that gnaws with sharp, chisel-shaped teeth. Most are mouse-like and vegetarian: they nibble plant stems, seeds and roots. Forty per cent of all mammal species are rodents. They live all over the world, from African jungles where crested rats clamber up trees, to scorching deserts where jerboas hop across the sand. House mice have even hitched lifts on ships to reach huts in Antarctica.

Dormouse

Moving House
Every three or four years, thousands of lemmings dash from their overcrowded homes. Many die in the frantic search for new places to live and feed.

A Plague of Rats
Every year, millions of nibbling rats wreck one fifth of the world's crops!

Many brown rats live in sewers. They use their feet as paddles when they swim and can tread water for three days!

Scaly tail

The greasy fur leaves dirty marks on things it touches.

Flat teeth at the back of a rat's mouth grind up grass and grain.

After walking through dirt, rats walk over food. This is how they spread diseases.

Straw nest

They are 18 days old and ready to leave home.

Their eyes begin to open and fur appears when they are ten days old.

The Most Common Mammal
House mice are small and cannot defend themselves. The species survives by having lots of babies. A female can give birth when she is six weeks old and have ten litters a year. If all her babies survived, she could have a million descendants when she was one year old!

Out on a Limb
Harvest mice climb up small plants, like monkeys in a tall tree! They hold on tightly with their tails.

Nonstop Teeth
Your adult teeth grow in the spaces left behind by your baby teeth. When they have filled the gap, they stop growing. Rodents' front teeth are different – they never stop growing! They get worn down by the tough food that the rodents eat.

Rodents even chew soap.

Growing tooth

Harvest mice weigh only five grams so they rarely break a 'branch'.

Tiny Nibblers

Pale kangaroo-mouse

Sagebrush vole

Great jerboa

Indian gerbil

Fat dormouse

Crested rat

LARGE RODENTS

Most rodents are small and look like mice, but some are much bigger. There are two sorts of large rodents – those that look like squirrels and those, such as porcupines, that look like pigs. Rodents' teeth get worn down by gnawing, and all rodents have front teeth that keep growing throughout their lives.

Chipmunks

Size of a Sheep!
The capybara is the largest rodent in the world. It is a relative of the guinea pig.

The Lion Lost
A porcupine will run backwards and stick its quills into an attacker's face!

A porcupine can hear fruit drop to the ground several metres away.

Hollow, stripy quill

The nest, or drey, is the size of a football.

Porcupines chew old bones to keep their teeth sharp.

Angry African porcupines stamp their back feet to rattle their quills. This warns other animals to go away.

Big Nibblers

The sharp quills only stick up when the porcupine is attacked.

European red squirrel

Chinchilla

Rock cavy

Spring hare

Woodchuck

Planting Trees
Squirrels bury nuts to eat later. Those that they can't find again may grow into trees.

A squirrel's strong jaws can crack open acorns.

Warm, leafy lining

Digging 'Dogs'
Prairie dogs live in underground towns! These towns usually have a population of about 1,000, but one in Texas had more than 400 million prairie dogs in it.

The 'garden' is weeded.

Watch 'dog'

Prairie dogs touch teeth when they meet.

Tufty Squirrel
European red squirrels don't scamper through gardens and parks like their grey American cousins. They are shy and hide from people in forests.

When it sleeps, the squirrel wraps its bushy tail around it, like a blanket.

Grey squirrels race through trees at up to 30 kilometres per hour.

A grey squirrel can leap more than six metres from one tree to another.

Squirrels can walk up and down the sides of trees to reach their nests.

A hole in the roof lets in air.

The house, or lodge, is the size of a large tent.

The underwater entrance keeps out enemies.

Food store

Timber!
American beavers are brilliant builders. They gnaw through trees with their teeth and then pile the logs up in a river. Behind this dam, a pond soon forms where they can build their home.

Dam

MARINE MAMMALS

Some mammals do not live on land, they swim in the sea. Sea cows, narwhals and all other marine mammals have to come to the surface to take in air – they do not have gills for breathing like fish. Walruses, sea lions, seals and sea otters can all climb out of the water. Sixty-five million years ago the ancestors of whales and dolphins could walk too. Skeletons of some whales still have leg bones!

Bowhead whale's 'leg'

A Mermaid?
Long ago, sailors mistook gentle seacows for magical, half-human creatures. They called them mermaids.

Blushing Blubber
When walruses sunbathe, blood rushes to the surface of their skin to cool them down and they turn pink!

Sea lions can twist their back flippers forwards.

This brown walrus has just left the icy water.

Five long toes, joined by skin, form a flipper.

California sea lions
grow up to 2.4 metres long. They live in the Pacific Ocean.

Underwater Birth
Baby blue whales are the biggest babies in the world! They are more than 1,000 times heavier than human babies.

The baby, or calf, rests on its mother's back and takes its first breath.

The playful calf drinks 600 litres of milk a day. It doubles its weight in the first week.

Whales are born tail-first to stop them drowning.

A sea lion's large eyes can only see in black and white.

When it dives, the sea lion closes its nostrils and blocks its windpipe with its tongue. This stops it swallowing water when it snaps up a squid, fish or octopus.

Ear flap

Sea lions bark!

Thick whiskers help the sea lion find food. They feel the ripples that fish make.

A sea lion's smooth body can speed through the sea at 40 kilometres per hour.

Waterproof, oily hair

Hope for the Future
Whales are so mysterious and massive that people will pay to see, or even stroke, a wild whale. Perhaps hunters will stop killing whales when they find they can make more money out of tourism.

The front flippers push the sea lion through the sea.

Sea 'Bird'
Beluga whales used to be called sea canaries because they chirp like birds. They can pull their lips into all sorts of shapes to squeal, whistle and 'moo' too!

A thick layer of fat, called blubber, keeps most marine mammals warm.

Mighty Milk
Seal's milk is full of fat and richer than the cow's milk you drink. Elephant seal pups put on over six kilograms every day!

This pup has a lot of growing to do – its father weighs more than two tonnes.

Father

Mother

THE HORSE FAMILY

Almost all the horses in the world are tame. The only wild species left is the Przewalski's horse. It survives in zoos. The commonest wild member of the horse family today is not a horse, but a zebra. Swift, stripy zebras are herbivores – they chew grass with their flat back teeth. Horses, zebras and asses walk on the tips of their toes, which are hidden inside hard 'shoes', called hooves.

Przewalski's horse

Desert 'Donkeys'
Wild asses are shy and rare. They live in the dry North African deserts.

A zebra's mane stands up straight.

The ears twist round to listen for danger.

Plains zebras are up to 1.2 metres tall at the shoulder. They live in Africa.

A zebra can see in colour during the day and as well as an owl at night!

Best Friends
Zebras and wildebeest like to live together. The zebras keep watch, while the wildebeest sniff the air for lions.

Wildebeest eat the short grass left by the zebras.

The hoof is just a large toenail!

Foals can walk when they are a few minutes old.

34

Quick, Run!
Asses, zebras and horses can run at speeds of up to 50 kilometres per hour.

All four hooves leave the ground when it gallops.

Long legs are best for taking big steps and running fast.

Plains zebras are plump.

Spot the Zebra!
Nobody knows why zebras are striped. Perhaps their crazy pattern puzzles lions, making it hard for them to pick their prey.

A zebra scratches its back by rolling on the ground.

Trespassers will be Kicked
Mountain zebras don't live in herds. Each male, or stallion, 'owns' a patch of land, called a territory. He will chase off any male zebra that enters his home.

Big thigh muscles power plains zebras across the African grasslands.

The swishing tail swats annoying, buzzing flies.

Females, foals and stallions live together in the same herd.

Make Your Mark
Your fingerprints are different to everybody else's. So are a zebra's stripes – each one wears a different patterned coat! Some even have thin white stripes on a black background.

There is only one toe inside this hard hoof.

Born Free
Some of the 75 million tame horses have escaped and learned to live in the wild. The ancestors of these American 'wild' horses, called mustangs, belonged to cowboys.

RHINOS AND TAPIRS

Thick Skinned
This rare Indian rhino has thick skin which protects it from spiky forest plants. Deep folds in its knobbly skin make it look as if it is wearing armour.

Rhinos and tapirs are related – they both have three toes on each foot, eat only plants and like to be alone. Apart from size, the main difference between these shy, quiet animals is that the rhino has a horn on the end of its nose. Rhino is short for rhinoceros, which means 'horned nose'. These ancient animals have lived on Earth for more than 55 million years but may not survive much longer. Rhinos are hunted and the tapir's forests are being cut down.

Rhinos cannot see things clearly if they are more than 30 metres away.

Muscles in this hump hold up the huge head.

White rhinos have a second, smaller horn.

The horn is made of hair, not bone. It can grow to be as long as a bath tub!

A wide mouth is good for picking big mouthfuls of grass.

Disappearing Rhinos
In 20 or 30 years' time there may be no rhinos left. They are being slaughtered so that their hairy horns can be made into dagger handles or crushed into a powder that is used as a medicine.

This four-week-old calf will stay with its mother until the next baby is born in two years' time.

Tiny hoof

Is it a Rhino?

Rhinos have changed shape and size over a long time. The first ones were the size of big dogs. Twenty million years ago they were taller than giraffes.

This rhino was the biggest mammal ever to live on land.

Early rhinos looked like tapirs.

During the ice age, 15,000 years ago, rhinos were woolly.

Charge!

Puffing like a steam train, black rhinos often charge at moving objects they don't like or recognize. As they are short-sighted, they gallop blindly at anything.

White rhinos are in fact grey, but they often roll in pale soil which makes them look white.

White rhinos are 1.8 metres tall. They live in Africa.

Female rhinos can produce 20 litres of milk a day.

A white rhino can weigh more than two tonnes. It is heavier and bigger than a family car.

Rhinos can sleep standing up.

Cat Food!

The donkey-sized Malayan tapir is the tiger's favourite food. To escape being eaten, this mother and baby are plunging into the water – they can swim better than cats.

Hide and Seek

Baby Brazilian tapirs have coats covered in yellowish spots and stripes. This pattern makes them hard to see in the light and shadows of their jungle home.

HIPPOS, PIGS AND PECCARIES

Hippo is short for hippopotamus, which means 'river horse'. They are called river horses because they live in rivers and lakes and eat grass. Pigs and peccaries do not live in rivers but they enjoy wallowing in mud as much as their huge relatives. Although these water-loving mammals are not carnivores, they are all able to protect themselves. Wild pigs can stab and kill tigers with their tusks, peccaries fight jaguars, and a heavy hippo will tussle with a crocodile or smash into a boat!

Very Important Pig
Farmyard pigs have all been bred from wild boar.

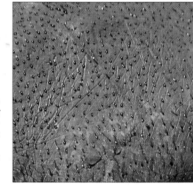

Built-In Suntan Lotion
Hippo skin oozes tiny blobs of pink liquid. This oil stops their skin from drying out and also protects them from sunburn.

Open Wide
You yawn when you are tired or bored, but male hippos yawn when they are angry! Smaller males are frightened off by the big teeth and swim away without starting a fight.

The eyes and nostrils are high up on a hippo's head. This means it can stick just the top of its head out of the water and still see and breathe.

Smooth, almost hairless, skin.

The half-metre-long tusks are sometimes used to stab crocodiles.

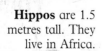

Hippos are 1.5 metres tall. They live in Africa.

This big male hippo weighs as much as 120 eight-year-old children!

Lumbering Lawn Mowers

Every night, hippos leave the water and spend five or six hours grazing. They troop back into the water, down well-worn paths, long before the scorching sun rises.

Plucky Peccary

If a mountain lion attacks a group of peccaries, one brave animal squeals loudly and runs towards the lion. This peccary dies, but the mothers and babies escape.

Piggy Parade

To keep cool and moist, hippos spend 16 hours a day up to their necks in water.

Underwater Ballet

Hippos can hold their breath for more than five minutes. This is plenty of time to dive down and tiptoe gracefully across the bottom of the lake.

Bush pig

Collared peccary

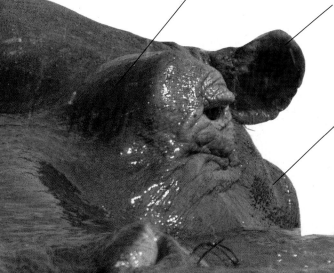

The hippo shuts its ears and nostrils when it is under water.

Thick skin protects the hippo from snapping crocodiles.

There are four toes on each foot.

Hippo

Hippos do eat water lilies, but prefer grass.

Pigmy hippo

THE CAMEL FAMILY

You find camels and their smaller South American relatives, vicuñas and guanacos, on sandy deserts, rocky plains and bare mountains. They survive in some of the harshest places on Earth. Vicuñas can breathe thin mountain air and camels can cope with freezing nights and scorching desert days. People have made good use of these animals' amazing survival skills – most are domesticated and work for a living. There are few wild members of the camel family left.

The fat inside a camel's humps shields its big body from the scorching sun.

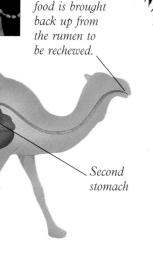

Domestic Dromedary
Desert peoples could not survive without their one-humped camels. They are ridden, milked and eaten. Camel skin is made into shoes, hair is woven into clothes and dry droppings are used as fuel!

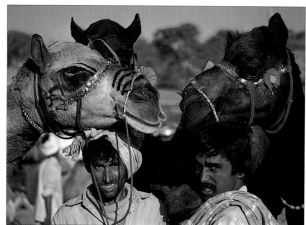

Partly digested food is brought back up from the rumen to be rechewed.

Chewed food travels first into the huge rumen.

Third stomach

Second stomach

Swallow That!
After going 10 months without water, a thirsty camel can gulp down nine big buckets of water in just 15 minutes.

Twice as Tasty
To get all the goodness out of grass, some mammals, such as camels, deer and cattle, have more than one stomach and chew their food twice!

The ears and nostrils can be pressed flat to keep out sand.

Two rows of eyelashes keep out sand and stop the eyes from freezing on cold desert nights.

A camel doesn't waste water. Liquid from its runny nose drips down the split lip into its mouth!

Bactrian camels are 2.2 metres tall. They live in the Gobi desert in Asia.

How Many Humps?

Dromedary

Camels spit at things that annoy them.

Tough lips can grip thorny desert plants.

Camels hardly ever sweat. This saves water.

Bactrian camel

King of the Castle
While the females graze, the male vicuña stands on a rock. If it spots a mountain lion it whistles and the fleecy females flee.

Guanaco

Camels roll from side to side when they walk because they lift both legs on one side at the same time.

Vicuña

The two toes spread out to stop the camel sinking into soft sand.

Hard-Working Mammals
Llamas and alpacas have been bred by people from wild guanacos. Llamas are milked and used to carry heavy loads. Alpacas are kept for their fine wool.

Alpaca

Llama

DEER AND GIRAFFES

First year

Third year

Giraffes, all male deer and female reindeer have spikes of solid bone growing out of the tops of their heads. They are used by the males to fight for females. The giraffe's bony stumps are covered in skin, but most deer have bigger, bony headgear, called antlers. These are covered in furry skin, or velvet, while they are growing, but they drop off and grow again every year. In just a few months, a moose can grow antlers that are bigger than you!

Male deer are called stags.

With eyes on the sides of its head, the deer can see all around and easily spot danger.

Male and female red deer only meet for a few weeks each year. They usually live in separate herds.

No Red Nose

Reindeer live in cold countries. They have thick fur to keep them warm – even their noses are furry! They find food under the snow by digging with their antlers and hooves.

Red deer live in Asia and Europe. Males are 1.3 metres tall at the shoulder.

Musk Maker

Male musk deer have a gland on their tummies that makes a smelly liquid, called musk. This long-lasting scent is an ingredient of many perfumes.

Wading In

In the winter, moose eat pine cones. In the summer, they wade into water and munch more than 1,000 salty water plants a day.

Fifth year

Seventh year

Growing Up
A red deer stag's antlers drop off every spring and are replaced by a new, bigger and better set. Its heavy headgear only takes about 100 days to grow.

Bony stump

Giraffes are the tallest mammals. Males are about five metres high.

For six weeks each autumn, males fight with their antlers. The older males, with larger antlers, win the females.

Legs Apart!
When a giraffe wants to drink from a pool, it has to stretch its long legs out wide – only then can it reach the ground.

Female red deer, called hinds, do not grow antlers.

This winter coat is thick and brown but a red deer's thin summer coat is a redder colour.

Red deer eat leaves and grass for ten to twelve hours a day.

Hoof

Hide and Spot
Baby deer, or fawns, have spotty coats for camouflage. They lie still in tall grass and hope that no deer-eating animals see them.

CATTLE AND ANTELOPES

Round Up
Musk oxen form a ring around their babies to protect them from hungry wolves – who can't pass the circle of horns.

Cattle, antelopes and their relatives goats and sheep, are all bovids. This means that they have horns firmly fixed to the tops of their heads. Horns have a bony core and an outer layer made of the same stuff as your fingernails. Gerenuk antelopes browse on leaves but bison prefer to graze on grass. Like all bovids, they take the goodness out of their poor-quality food by coughing up partly digested food and chewing it a second time. This is called chewing the cud.

Gerenuk (male)

All cattle have four stomachs!

Arabian oryx

These horns are about half as tall as you!

This rare Arabian oryx antelope chews the cud as it walks across the desert.

On the March
At the onset of the dry season, huge herds of wildebeest walk over 1,600 kilometres to wetter, greener pastures. When the wet season begins, they wander back. These long, yearly journeys are called migrations.

Like all cattle, bison have split hooves.

It sniffs the air for rain and then walks to where the grass is growing.

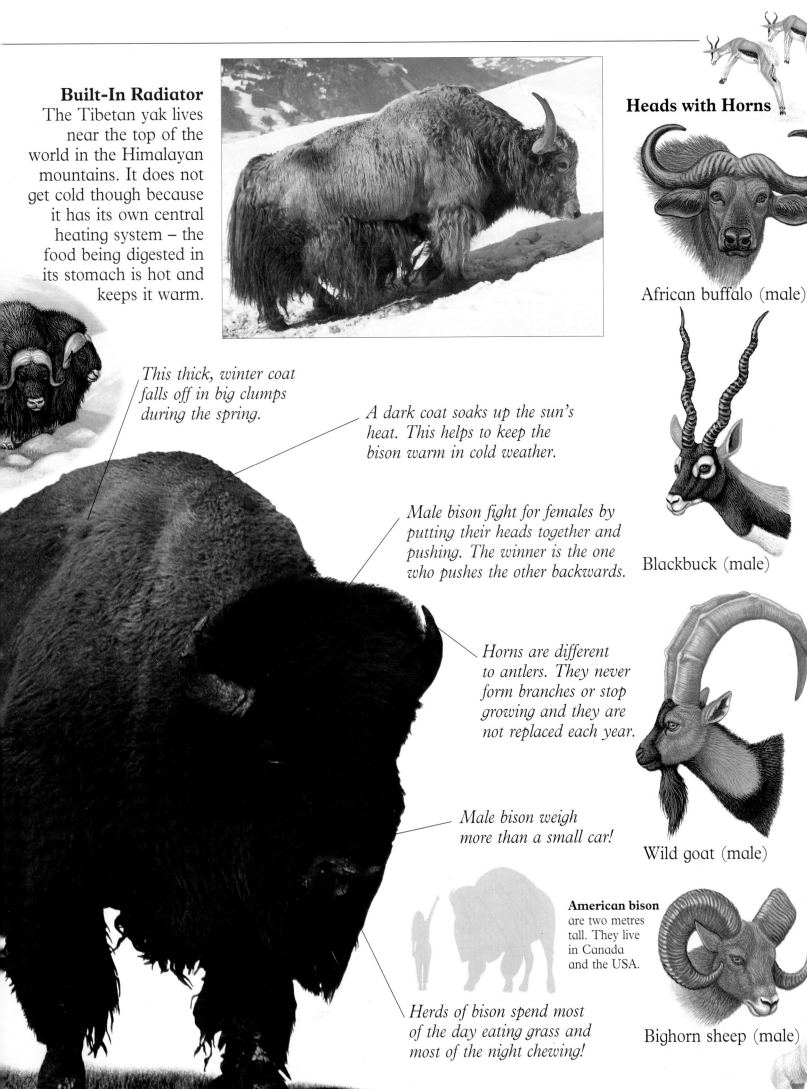

Built-In Radiator
The Tibetan yak lives near the top of the world in the Himalayan mountains. It does not get cold though because it has its own central heating system – the food being digested in its stomach is hot and keeps it warm.

Heads with Horns

African buffalo (male)

This thick, winter coat falls off in big clumps during the spring.

A dark coat soaks up the sun's heat. This helps to keep the bison warm in cold weather.

Male bison fight for females by putting their heads together and pushing. The winner is the one who pushes the other backwards.

Blackbuck (male)

Horns are different to antlers. They never form branches or stop growing and they are not replaced each year.

Male bison weigh more than a small car!

Wild goat (male)

American bison are two metres tall. They live in Canada and the USA.

Herds of bison spend most of the day eating grass and most of the night chewing!

Bighorn sheep (male)

ELEPHANTS

Elephants have enormous ears, long noses, tusks – and weigh more than six cars. They are the biggest land mammals. Herds of elephants shape the land they live in by treading paths that are wide enough to stop bush fires, by digging wells in dry riverbeds, by fertilizing the ground with dung and by trampling grass for zebras to eat. They also open up forests by pushing over trees!

A Family of Females
The leader of a herd of elephants is an old female, called a matriarch. She is followed by all her female relatives and their babies.

Wrinkles trap water and help to keep it cool.

Ivory Towers
Many elephants are shot for their valuable ivory tusks. People have burnt huge piles of old tusks to show that they want this cruelty to stop.

Elephants never stop growing.

It's all Relative
The elephant's closest relative, the hyrax, looks like a guinea pig! Like elephants, hyraxes have nails instead of hooves. They also have long incisors, like the elephant's tusks. They live in Africa.

Tusks are teeth. They grow about 17 centimetres a year and can be as long as a car!

Who's Who?
There are two sorts of elephant.

Ankle

Humped back

Smaller ears

Longer tusks

Bigger, rounder ears

Taller

Asian

African

Elephants eat grass, bark and leaves for up to 20 hours a day.

This toenail is bigger than your whole hand!

Young females
act as nannies.

An African
elephant's ears
are almost as
big as sheets for
a single bed!

Males leave their families
when they are about fourteen.

Stay Cool
Elephants have lots of
ways of cooling their big
bodies. They can wallow
in mud or throw water
and dust over their skin.
Sometimes they flap their
ears like giant fans!

The elephant spreads
out its ears to make
itself look bigger and
more dangerous.

Elephants keep in
touch by making
deep, rumbling noises
in their tummies!

The bendy trunk is formed
from the nose and the upper
lip. It is used for breathing,
smelling, touching and
picking things up.

Elephants can live to celebrate
their eightieth birthdays!

This is one of
the first four teeth.

The sixth, and
last, set of teeth are
bigger than bricks!

What's Inside?
Elephants may
look as if they have
flat feet but they
really walk on
their tiptoes!

Toe

The heel rests on
a fatty cushion.

Always Teething!
Apart from tusks, elephants have only four
teeth. These molars are replaced every few
years. Bigger teeth appear at the back of
the mouth and push out the old, worn
teeth – like a conveyor belt of teeth!

GLOSSARY

Antler Solid bone that grows out of the top of a mammal's head. Antlers fall off and regrow every year.

Browsing Eating the leaves and twigs of trees and small bushes. Giraffes browse.

Camouflage Special patterns and colours on an animal's coat that help it to hide from danger.

Canine teeth Pointed dagger-like teeth that carnivores, such as tigers, badgers and wolves use to grip and kill other animals.

Carnivore An animal that mainly eats other animals.

Digesting Taking the goodness out of food to make energy.

Domesticated Kept and looked after by people and not living in the wild. Many horses and camels have been domesticated.

Extinct Animals that no longer exist or have not been seen for over 50 years.

Gnawing The constant chewing carried out by rodents to stop their teeth from growing too long.

Grazing Eating grass, usually by biting it down to the ground. Cattle graze.

Grooming Cleaning or combing fur. Many animals groom one another.

Herbivore An animal that eats plants, but not meat.

Herd A group of hoofed animals, such as bison, that feed or move together.

Hoof A covering of horn at the end of some animals' feet. Animals that have hooves, such as horses and rhinos, are called ungulates.

Horn Bone covered in horn that grows out of the tops of the heads of antelope and cattle. Horns do not fall off.

Mammary gland The soft part of a female mammal's body that produces milk. Babies usually suck this milk through a nipple.

Marsupial A mammal that is only partly formed when it is born and then carries on growing in its mother's pouch.

Milk A very rich drink that is full of goodness. All baby mammals are fed milk by their mothers.

Molar Large back tooth used for chewing. A zebra eats grass so it has a mouthful of molars.

Molehill A mound of soil that has been piled up above a mole's underground nest.

Moulting The regular shedding of one coat of hairs for another. Many mammals, such as camels, have warm winter coats and thinner summer coats.

Omnivore An animal that eats both plants and meat. All bears, except polar bears, are omnivores.

Pack A group of animals that survive by working together. Dogs hunt in packs.

Prehensile Able to grasp on to things. Some mammals have prehensile tails.

Primate A group of mammals that includes apes, monkeys and people. Primates have hands that can grasp and large brains.

Rumen A special part of the stomach in which food is partly digested, before being brought back up to the mouth to be chewed again.

Scavenger An animal that eats meat but that does not usually kill other animals. Scavengers eat other animals' leftovers.

Species A type of animal. Gorillas and hippos are different species, so they cannot have babies together.

Suckling A baby mammal feeding on its mother's milk.

Territory The patch of land or water where an animal lives and feeds, and which it defends against other animals.

Trunk A long, bendy part of an elephant's body made up from the top lip and nose. It can pick things up.

Tusk A large tooth. Elephants have long tusks.

Acknowledgments

Photography: Andy Crawford, Dave King, Steve Gorton, Tim Ridley, David Rudkin, Harry Taylor and Jerry Young.

Illustrations: Roy Flooks, Ray Hutchins, Stuart Lafford, Kenneth Lilly, Steve Lings, Sean Milne, Richard Orr, Brian Watson and Phil Weare.

Model: Donks Models.

Thanks to: Penny Boyd at Burstow Animal Wildlife Sanctuary; Caroline Brooke; The Colour Company; Tina Lewis; Natural History Museum, London; Norrie Carr Model Agency; Scallywags Child Model Agency; Richard Walker; Whipsnade Zoo; Julie Whittaker.

Picture credits

Heather Angel: 43tr; **Ardea:** Jean-Paul Ferrero 34tr, Kenneth W. Fink 35cra, 37cr, Francois Gohier 41cb, Clem Haagner 11cl, J.M. Labat 8c, R.F. Porter 36bl, Adrian Warren 23b, 27cr; **Bruce Coleman Ltd.:** David Austen 40c, Jane Burton 18/19tc, 28cr, Alain Compost 22/23c, Peter Davey 46cl, Adrian Davies 24c, Francisco Erize 30tr, 36tr, Jeff Foott 35ca, Steven C. Kaufman 3c, 25bra, Stephen Krasemann 26br, William S. Paton 1c, 17cra, M.R. Phicton 38cr, Dieter & Mary Plage 47cr, Hans Reinhard 38tr, 45tc, Leonard Lee Rue 43br, Norman Tomalin 16cr, John Uisser 29cl, Konrad Wothe 25tl; **Robert Harding Picture Library:** 7br, 42cl; **The Image Bank:** Guido Alberto Rossi 40tc, Jack Ward 35bra; **Frank Lane Picture Library:** Eric & David Hosking 13tl, M. Macri 6c, W. Wisniewski 32clb; **NHPA:** Agence Nature 44c, Henry Ausloos 3tr, 38clb, Stephen Dalton Jacket, 5tlb, 12tr, 12clb, 28c, 28cla, Michael Leech 29l, Lacz Lemoine 34tl, Tsureo Nakamura 33br, Haroldo Palo 37br, S. Robinson 22c, 37tl, Jany Sauvanet 10/11tr, John Shaw 20cr; **Nature Photographers Ltd.:** Paul Sterry 33cra; **Oxford Scientific Films Ltd.:** Animals Animals/Breck P. Kent 44/45c, Hans & Judy Beste 26bl, Michael Fogden 27tl, Renee Lynn 15tr, Joe & Carol McDonald 8/9c, Tom McHugh 7tl, Stan Osolinski 16cra; **Planet Earth Pictures:** K. Ammann 38/39bc, Gary Bell 8/9cb John Bracegirdle 31cr, Jim Brandenburg 18cla, 20tr, Mary Clay 16cl, Richard Coomber 47tc, Ken Lucas 22cl, Richard Matthews 25c, Scott McKinley 42br, Doug Perrine 32/33ca, Ronald S. Rogoff 15c, 19trb, Jonathan Scott 46c, 46/47c, Anup & Manoj Shah 16bc, Peter Stephenson 44bl; **Premaphotos Wildlife:** K.G. Preston-Mafham 6cr; **Survival Anglia Ltd.:** Jeff Foott 39tr, Nick Gordon 15crb, J.M. Pearson bla, Alan Root 9cr, Alan & Jane Root 39cr, Vivek Sinha 42bl; **Zefa:** 22bl, F. Lanting 43c.

l – **left**   c – **centre**   b – **bottom**   tl – **top left**   cl – **centre left**   bl – **bottom left**   tr – **top right**
cr – **centre right**   br – **bottom right**   tc – **top centre**   bc – **bottom centre**   cb – **centre bottom**
ca – **centre above**   tlb – **top left below**   trb – **top right below**   cla – **centre left above**
clb – **centre left below**   cra – **centre right above**   crb – **centre right below**   bra – **bottom right above**

INDEX